ELEMENTS OF THE PERIODIC TABLE

WHAT IS CARBON?

006

C

CARBON

12.01

KATHLEEN A. KLATTE

ROSEN PUBLISHING

Published in 2026 by The Rosen Publishing Group, Inc.
2544 Clinton Street, Buffalo, NY 14224

Portions of this work were originally authored by Linda Saucerman and published as *Carbon*. All new material this edition authored by Kathleen A. Klatte.

Designer: Rachel Rising
Editor: Kathleen A. Klatte

Cataloging-in-Publication Data
Names: Klatte, Kathleen A.
Title: What is carbon? / Kathleen A. Klatte.
Description: Buffalo, NY : Rosen Publishing, 2026. | Series: Elements of the Periodic Table | Includes glossary and index.
Identifiers: ISBN 9781499478365 (pbk.) | ISBN 9781499478372 (library bound) | ISBN 9781499478389 (ebook)
Subjects: LCSH: Carbon--Juvenile literature. | Chemical elements--Juvenile literature.
Classification: LCC QD181.C1 K538 2026 | DDC 546'.681--dc23

Manufactured in the United States of America

CPSIA Compliance Information: Batch #CSRYA26. For further information, contact Rosen Publishing at 1-800-237-9932.

CONTENTS

INTRODUCTION 4

CHAPTER ONE
Blazing Through History 6

CHAPTER TWO
Setting the Table 14

CHAPTER THREE
From Diamonds to Pencil Shavings 22

CHAPTER FOUR
Life Is Like a Slice of Pizza 30

CHAPTER FIVE
Calling All Compounds 38

CHAPTER SIX
All About Isotopes 46

THE PERIODIC TABLE OF ELEMENTS 54

GLOSSARY 56
FOR MORE INFORMATION 58
FOR FURTHER READING 60
INDEX 62
ABOUT THE AUTHOR 64

INTRODUCTION

What element is found in everything that's ever lived on Earth? What do dinosaurs and diamonds have in common? What element can be both one of the hardest and softest substances known to science? What element can naturally exist as a solid or gas?

The answer to all these questions is carbon. The chemical symbol for carbon is C.

Carbon is part of everything that lives—or ever has lived—on planet Earth, including dinosaurs, trees, and people. Diamonds, one of the hardest natural substances known to science, are composed of carbon. They're also some of the oldest objects you're likely to encounter. Most natural diamonds were formed millions of years ago—longer than humans have existed!

The center of your pencil is probably graphite, the softest form of carbon. Carbon can also exist as a gas. It's probably most familiar to you as carbon dioxide (CO_2), which you exhale when you breathe. Carbon moves through the earth, air, oceans, and all living things. Plants take in carbon dioxide to make food. Animals ingest carbon in their food and release it when they breathe. When plants and animals die, they decompose, which releases carbon into the environment. These are just a few of the ways carbon molecules move through the world. This is called the carbon cycle.

The carbon cycle is very complex. Carbon molecules combine with other elements and move between the earth, air, and living organisms.

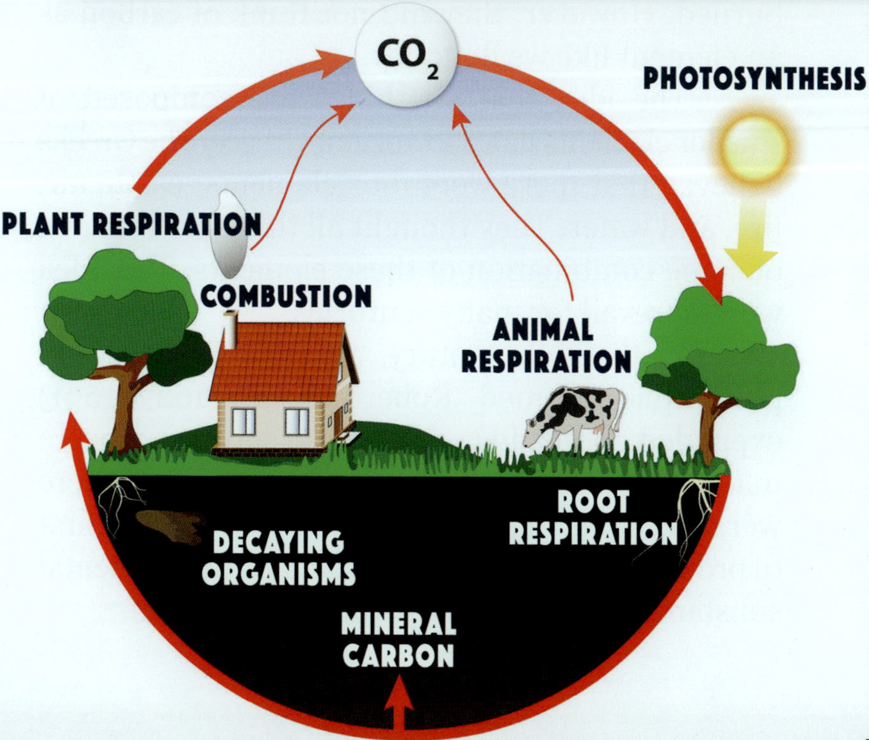

CHAPTER 1
BLAZING THROUGH HISTORY

The word "carbon" was first used in 1788. It comes from the French word *carbone*, which comes from the Latin word *carbo*, meaning ember or charcoal. Ancient people probably first encountered carbon in the remnants of fire—soot, ash, and charcoal. These substances are produced when another natural substance which contains carbon, such as wood, is burned. However, they did not think of carbon as an element like we do today.

The idea that all things are composed of smaller elements dates to ancient Greece. The Greeks believed that there were four elements: earth, air, fire, and water. They thought all things were made of some combination of these elements. This idea would prevail for many centuries.

In the 17th century, an Irish chemist and philosopher named Robert Boyle (1627–1691) expanded on this idea. He agreed substances were made of smaller particles, but he suggested there were more than just four elements. He was the first to propose the idea that an element is a fundamental substance made up of only one kind of atom.

Boyle was also one of the first scientists of his time to try to prove his theories in a scientific way. He performed experiments, keeping detailed records of what tools and materials he used and what results were observed. Before this, scientific theories were often discussed and debated more like philosophical questions.

French chemist Antoine Lavoisier (1743–1794) worked with carbon and gave it its name. Another scientist to work with the element was Sir Humphry Davy (1778–1829). Davy is noted for establishing the science of electrochemistry. He discovered the elements sodium, potassium, and chlorine using batteries. He also demonstrated that coal, soot, diamonds, and charcoal were all forms of carbon.

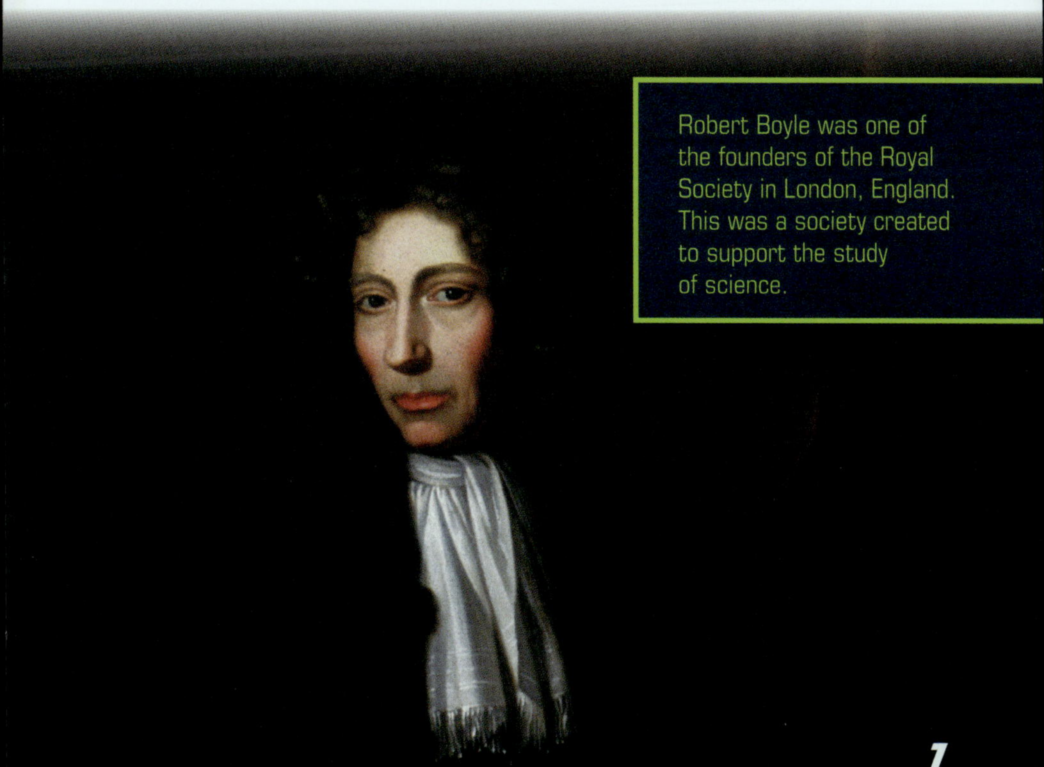

Robert Boyle was one of the founders of the Royal Society in London, England. This was a society created to support the study of science.

IT'S ELEMENTAL!

To understand how elements behave, you must think small—really, really small. An element is one of more than 100 known substances that cannot be separated into simpler substances by physical or chemical means. For example, if you take the element iron (Fe) and melt it down, you still can't turn it into another element. If you hammer a piece of iron, you also can't change it into another element. No two elements are exactly alike, and all elements are made up of atoms. Atoms are the basic building blocks of all matter. Atoms unite with other atoms to form molecules. Different forms of carbon, such as graphite or diamond, are made by different combinations of carbon atoms that make different molecules.

Carbon's entry in the periodic table of elements contains lots of information. Carbon's chemical symbol is the letter C. Its atomic weight, or mass—for the most part, the number of protons and neutrons it contains—is 12.011.

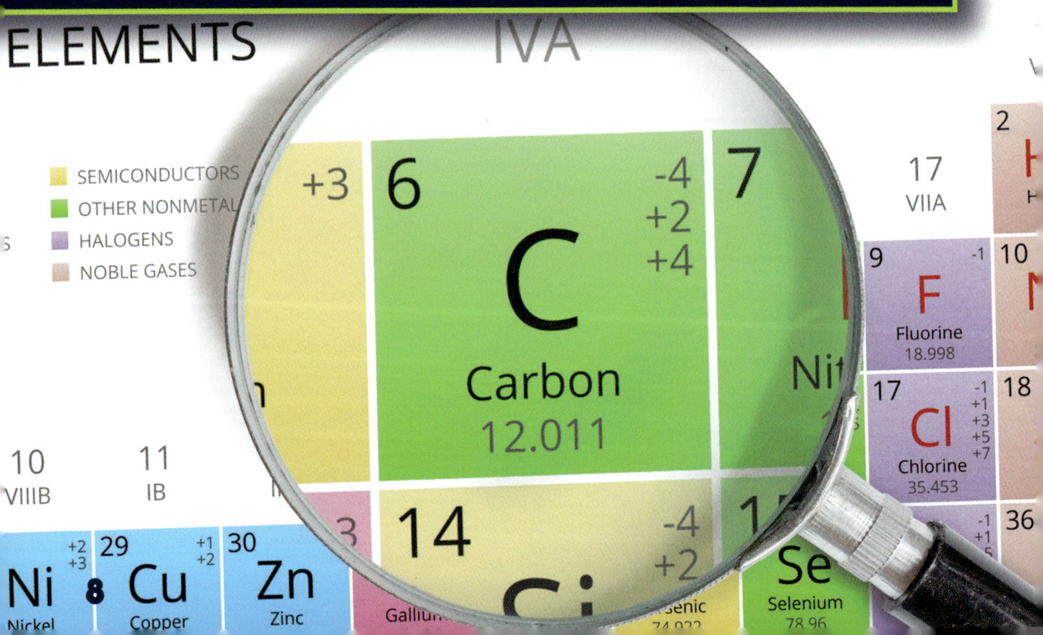

ATOMIC NUMBER
6

ATOMIC MASS
12.011

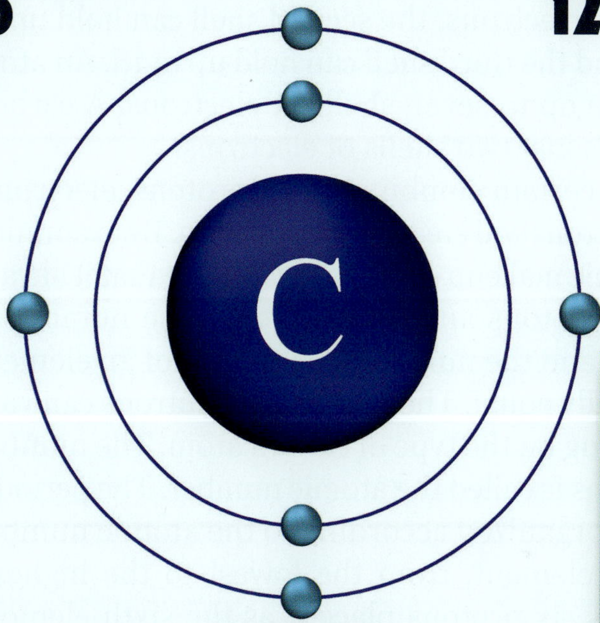

CARBON

ELECTRON CONFIGURATION : $1s^2$. $2s^2$. $2p^2$
ENERGY LEVEL : 2.4

Carbon's energy level reflects how many electrons are in the shells that are around the atom's nucleus. Its electron configuration describes where those electrons are found.

WHAT'S SMALLER THAN AN ATOM?

Until the late 19th century, scientists believed the atom was the smallest structure known to mankind. Then scientists learned that atoms have pieces too. The main particles that make up atoms are protons, neutrons, and electrons. These are called subatomic particles. Protons have a positive electric charge, neutrons have no charge, and electrons have a negative electric charge. All atoms contain a nucleus, or center, where the protons and neutrons are contained. Outside the nucleus are shells, or layers, which contain the electrons. The first shell can only hold two electrons, the second shell can hold up to eight, and the third shell can hold up to 18. An atom can have up to seven shells of electrons. A carbon atom has just two shells of electrons.

A certain combination of protons, electrons, and neutrons will create a carbon atom. By examining the atomic makeup of carbon, you'll see that it always has six protons and six electrons. The number of protons and the number of electrons of an element are usually equal. The number of neutrons can vary depending on the type of carbon atom. The number of protons is called the atomic number. The periodic table is organized according to the atomic number of each element, from the lowest to the highest. Carbon's six protons place it as the sixth element on the periodic table, so it is not a heavy element compared with the more than 100 elements that follow it on the table.

A proton is a positively charged subatomic particle. Elements are defined by the number of protons in the nucleus of an atom.

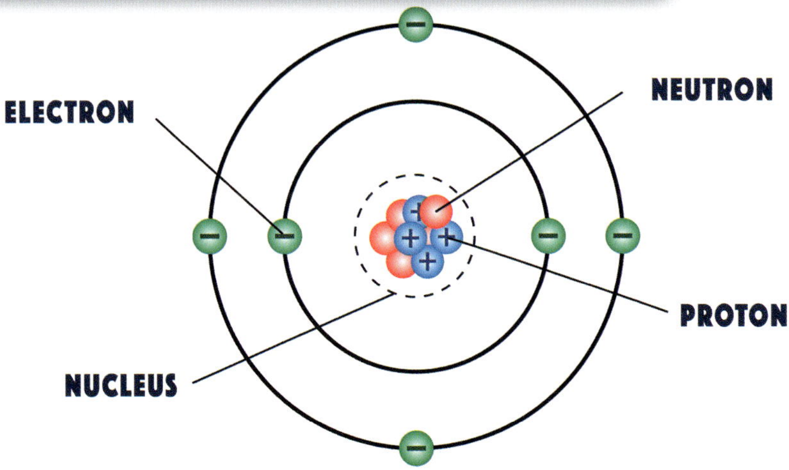

C

CHEMICAL SYMBOL: C

PROPERTIES: Nonmetal, combustible solid; comes in four allotropes: diamond, graphite, white carbon, fullerene

DISCOVERED BY: Not attributed to one person; element has been known of since prehistoric times

ATOMIC NUMBER: 6 **ATOMIC WEIGHT:** 12.011

PROTONS: 6 **ELECTRONS:** 6 **NEUTRONS:** 6

DENSITY OF SOLID: Diamond: 3.5 grams per cubic centimeters; Graphite: 2.2 g/c3

MELTING POINT: 6,332°F (3,500°C), 3,773 K

BOILING POINT: 8,721°F (4,827°C), 5,100 K

COMMONLY FOUND: Everywhere

Prehistoric humans first encountered carbon in the form of charcoal.

Carbon is noted for its ability to combine easily with other elements to form compounds. Carbon compounds make up nearly 20 percent of the matter of all life on earth. Carbon compounds are generally formed by two types of bonds. Covalent bonds are formed when electrons are shared between atoms. Ionic bonds are formed when one atom gives an electron to a different atom.

A carbon dioxide molecule is formed when one molecule of carbon forms covalent bonds with two molecules of oxygen, sharing four electrons.

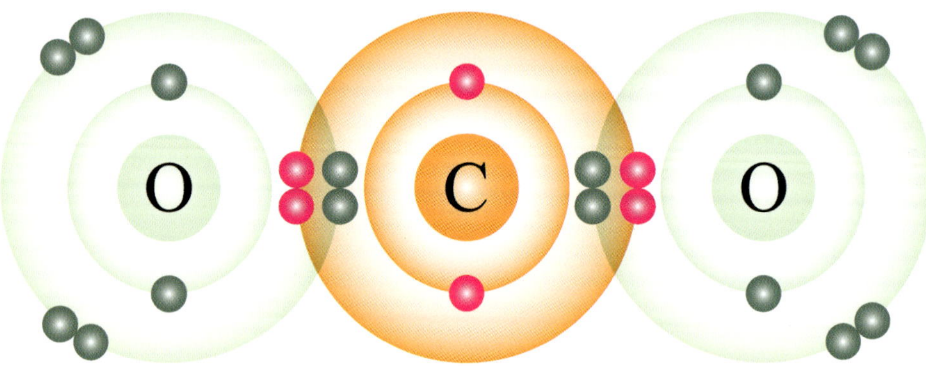

CARBON DIOXIDE

CO_2 : O=C=O

SHARING 4 ELECTRONS

Carbon's entry on the periodic table can tell us a lot about it. You just need to know what the numbers and letters mean!

CHAPTER 2

SETTING THE TABLE

The periodic table of elements is a staple in every chemistry textbook, classroom, and lab. But who created it? Why? How did they get the idea? The answers may surprise you.

The periodic table was designed by a Russian chemist named Dmitri Mendeleev (1834–1907). When he began teaching inorganic chemistry, he couldn't find a suitable textbook for his students. There was a great deal of information known about elements, but no clear way of organizing it. About 1867, he began writing a new textbook. It was called *The Principles of Chemistry*. As part of his research, he noticed the progression of atomic weights in groups of elements. He devised the idea of arranging elements in a table in order of their atomic weight. He also realized that there were probably more elements that hadn't been discovered yet and left room for them in his table.

Mendeleev's first version of the periodic table was published in 1869. It began with the lightest element, hydrogen (H), and went up to uranium (U), the heaviest known at the time. During Mendeleev's lifetime, he proposed changes to the periodic table based on updated research and the discovery of new elements that fit into the blank spaces in his table.

The modern periodic table takes Mendeleev's idea and makes it a little easier to understand. The periodic table we use today is arranged in horizontal rows called periods. In each period, the elements are arranged by atomic number.

Mendeleev's periodic table is shown on the Bureau of Weights and Measures in St. Petersburg, Russia. Mendeleev ended his career as head of the bureau.

Elements also are arranged in the periodic table by groups. Group numbers appear above each column of the table. There are two main subgroups, A and B, each of which has eight groups. For example, some group numbers are IA, IIA, IB, and IIB. All the elements within a group have similar chemical properties. These properties affect the way an element reacts during a chemical change (how the atoms are rearranged to make an entirely new substance). These differing chemical properties are sometimes referred to as families of elements. The elements in the A groups, such as carbon, are known as representative elements, or elements that add electrons to the outermost shell. Elements in the B group are transition elements, or elements that add electrons to the inner shells of the atom. Transition elements are malleable, conduct electricity and heat, and can form covalent bonds.

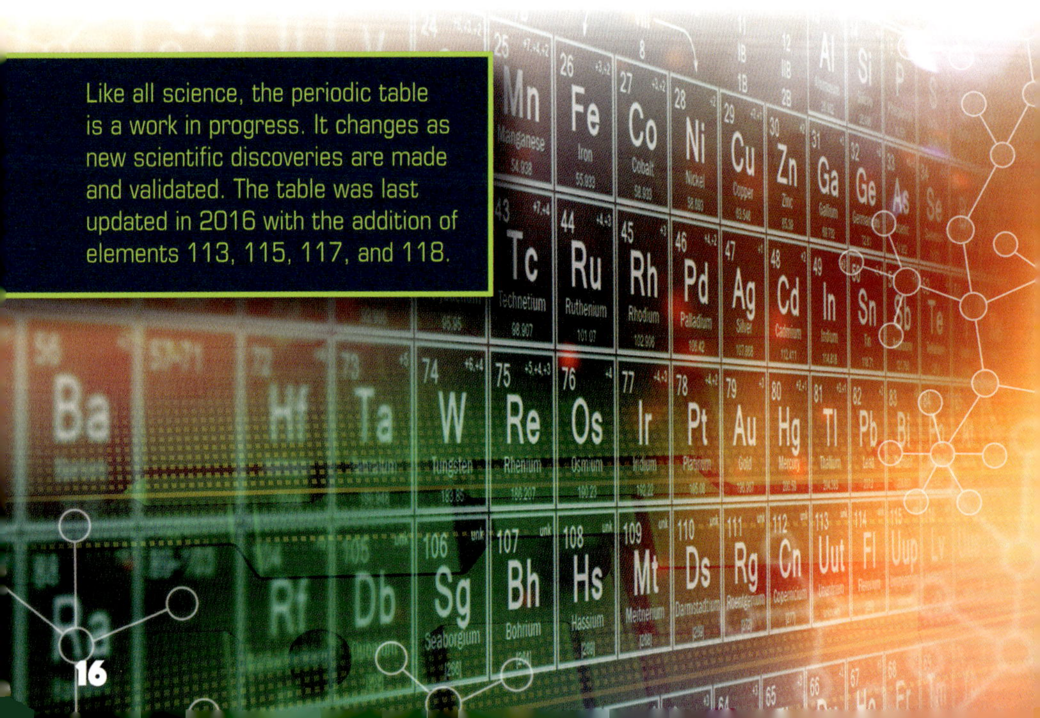

Like all science, the periodic table is a work in progress. It changes as new scientific discoveries are made and validated. The table was last updated in 2016 with the addition of elements 113, 115, 117, and 118.

Many of the groups have a name. For example, elements in group IVA, where carbon is the first element, are called the carbon group. Carbon is a nonmetallic element, which means that it doesn't act as a metal. However, the other elements in the carbon group—silicon (Si), germanium (Ge), tin (Sn), lead (Pb), and flerovium (Fl)—are not in the nonmetallic family. In fact, from silicon to flerovium, these elements are each a metal in some form. However, all share some qualities, which is why they are in the same group.

LET'S BE FRIENDS!

Elements in certain groups have similar chemical behaviors because of their electric forces. If you've ever held two magnets and felt the strong force that makes it impossible for you to stick the two magnets together, then you have an idea of what happens when two protons are near each other. Because protons are positively charged, they repel each other. Neutrons, which are electrically neutral, can help lessen this repulsion between protons. But it is the negatively charged electrons that really determine how elements behave and react with one another.

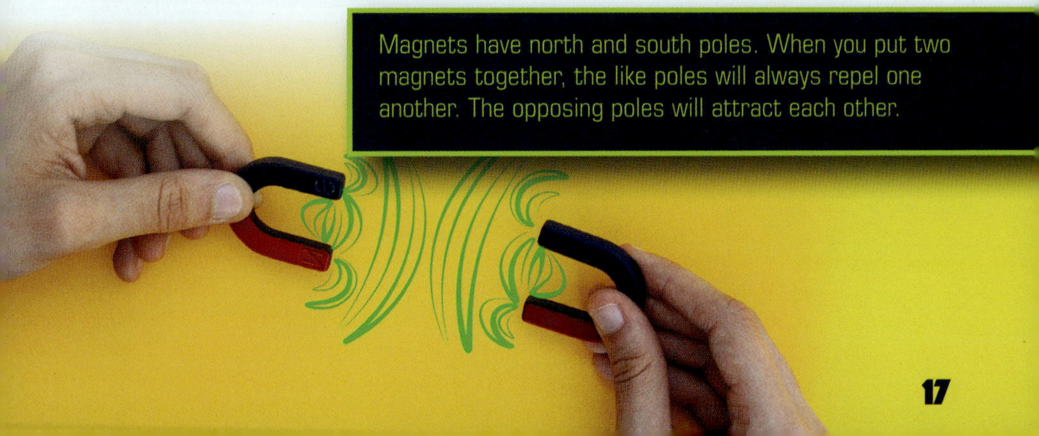

Magnets have north and south poles. When you put two magnets together, the like poles will always repel one another. The opposing poles will attract each other.

In a carbon atom, four of the six electrons are in the outermost shell of the atom and are known as valence electrons. These valence electrons are very "friendly," as they allow carbon to form covalent bonds. Covalent bonds result when two atoms share a pair of electrons. For instance, the air we exhale from our lungs is carbon dioxide (CO_2)—the combination of one carbon atom and two oxygen atoms sharing electrons. In addition to bonding with other elements, these valence electrons allow carbon to bond with itself to create other substances like diamonds and graphite.

Not every element occurs in nature. Some are created in laboratories. This includes some of the elements in the carbon group.

HEAVYWEIGHTS

The latest additions to the periodic table are four superheavy elements created in laboratories. In 2016, elements 113, 115, 117, and 118 were added to the table.

ATOMIC NUMBER	INITIAL DESIGNATION	OFFICIAL NAME
113	UNUNTRIUM (UUT)	NIHONIUM (NH)
115	UNUNPENTIUM (UUP)	MOSCOVIUM (MC)
117	UNUNSEPTIUM (UUS)	TENNESSINE (TS)
118	UNUNOCTIUM (UUO)	OGANESSON (OG)

At this time, most scientists believe that all naturally occurring elements have been discovered.

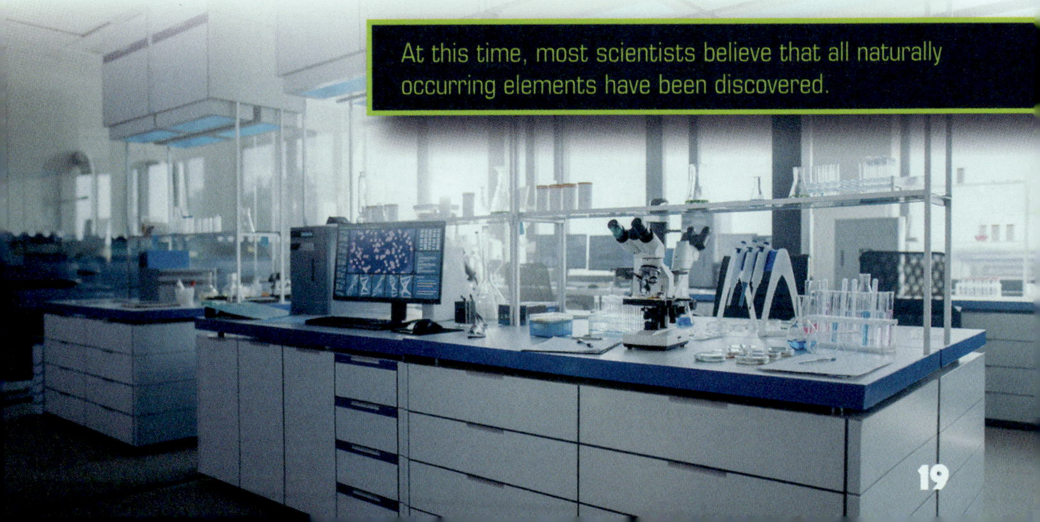

Flerovium is the final element in the carbon group. It's a transuranium element, meaning it has an atomic number greater than uranium, which has an atomic number of 92.

The final element of the carbon group was discovered in 1999 by scientists in Russia and California who were experimenting with radioactive isotopes. An isotope is an atom that has the same number of protons as usual but a different number of neutrons. Initially, the element was named ununquadium. It has an atomic number of 114.

The discovery of ununquadium was officially recognized by International Union of Pure and Applied Chemistry (IUPAC) and the International Union of Pure and Applied Physics (IUPAP) in 2011. It was renamed flerovium.

Flerovium is named for prominent Russian physicist Georgy Flerov (1913–1990).

CHAPTER 3

FROM DIAMONDS TO PENCIL SHAVINGS

Carbon exists in many forms—including the hard, valuable diamonds in jewelry stores and museums and the soft graphite shavings in your pencil sharpener. The different physical forms an element can take are called allotropes. Some allotropes have distinct crystal forms; some don't.

Allotropes are possible because of the way atoms are structured. Electrons orbit the nucleus of an atom in shells. Carbon contains two shells of electrons. The outer shell is called the valence shell. The valence shell of carbon contains four electrons. However, it has room for eight. Carbon can share or acquire these extra electrons from other atoms. This allows it to exist in such different forms.

ALL THAT GLITTERS

In diamonds, each carbon atom shares all four of its outer electrons with other carbon atoms through a series of bonds. These bonds form the diamond's unit cell, which is the unit repeated to form a crystal. A crystal is a material that takes on a shape according to a repeating structure of unit cells. The tightness between the atoms in the crystals contributes to a diamond's hardness. It is also why diamonds have such a high melting point—about 6,332°F (3,500°C) to 8,132°F (4,500°C)!

The arrangement of the crystals determines a diamond's shape. An octahedron, an eight-sided figure, is the most common shape for a diamond. A diamond can also take on other shapes, depending on how the bonds are repeated. The tight bonds and density of diamonds give a clue to their origins. To create a diamond, carbon atoms have to be squeezed together under an enormous amount of pressure and in high temperatures: about 725,000 pounds per square inch (50,972 kg per sq cm) and about 2,200°F (1,204°C).

Naturally occurring diamonds form deep in the earth, in an area called the mantle where heat and pressure are intense. Diamonds are common in South Africa, where they can be found in ancient volcanoes, in riverbeds, and on the ocean floor. Diamonds are also found in Australia, Brazil, Canada, Russia, and other countries. Microscopic diamonds have also been found in some meteorites. Natural diamonds can be clear, cloudy, or a variety of colors including pink, purple, yellow.

Carbon allotropes have unique structures. Diamonds form tight bonds, which form hard crystals. Graphite has loose bonds that break easily. And fullerenes have a distinctive round shape that scientists are still learning about.

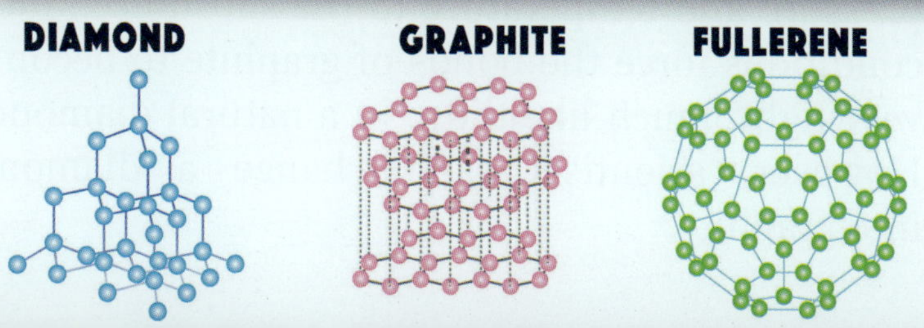

Some of the world's most famous diamonds, including the Hope Diamond and the Koh-i-Noor, originated in India. India was the world's only source of diamonds until they were discovered in Brazil in 1725.

WHAT'S IN YOUR PENCIL?

On the other end of the spectrum is another form of carbon called graphite. Graphite is a form of carbon that is dark colored, has a layered structure, and is very soft. Unlike those in a diamond, the carbon bonds of graphite are very loose. The loose bonds allow graphite to move very easily and create a substance with an often-hexagonal (six-sided) crystal shape. When graphite is used in a pencil, its loose bonds break apart, easily transferring onto a piece of paper as you write.

In the 1950s, scientists discovered how to make diamonds out of graphite by applying very high pressure and temperature to the substance. Such conditions force the bonds of graphite to become very tight, much like those in a natural diamond. However, scientists can't change a diamond into graphite.

Artificial diamonds aren't sought after as much as the naturally occurring gems. But science has improved artificial diamonds so much that it is difficult for most people to tell the difference between a natural and a man-made stone.

Many synthetic diamonds are used as part of tools. These diamonds have the same hardness as natural diamonds, but they can be larger and are used to cut concrete and bricks and to drill deep into the earth.

Like natural diamonds, synthetic diamonds have an extremely high rate of thermal conductivity. This means that a tremendous amount of heat can pass through a diamond without damaging it. Because of this, scientists are looking at synthetic diamonds for use as semiconductors, which are materials used to transmit electricity.

The loose bond between the carbon atoms that form graphite is what makes it such a soft mineral.

A pencil's lead, or core, is probably the most familiar form of graphite. When you press your pencil to paper, the loose bonds between the carbon atoms break, leaving marks on the paper.

CHAOITE

Another possible allotrope of carbon is called white carbon or chaoite. It was discovered in 1969 and is thought to be formed during meteorite impacts. Not much is known about white carbon, and scientists are still considering whether it is a true allotrope.

FULLERENES

In 1985, scientists discovered a fourth allotrope of carbon, fullerenes. Buckminsterfullerene (C_{60}), which is the most common form of this allotrope, was produced when a team of scientists at Rice University zapped graphite with laser beams. In 1996, they won the Nobel Prize for their work.

The carbon that was formed had a structure that looked like the design of a soccer ball. The scientists named the new allotrope after the engineer R. Buckminster Fuller, who designed dome houses that resembled soccer balls, giving the allotrope the nickname "buckyballs." There are also fullerenes with a structure that looks like a cylinder or tube.

In 2014, researchers in Moscow used fullerenes to produce a substance even harder than diamonds.

R. Buckminster Fuller (1895–1983) was an American inventor and architect. He held numerous patents and honorary degrees and authored 28 books. His most famous creation is the geodesic dome.

OTHER FORMS OF CARBON

Amorphous carbon is another form of carbon that has no crystalline form. Many things may fit in this category, and the carbon can take many different shapes.

Some impure forms of carbon are often called amorphous, although they're not. These forms include charcoal, coal, and soot.

Coke is made by heating coal in the absence of air. The fire from heating coke will burn hotter than a wood fire and has been used to recover iron from ore, the natural mineral from which the metal is extracted. Fossil fuels, including coal, oil, and natural gas, are mostly carbon and can be burned to create heat or power.

The earliest mention of diamonds dates back to the fourth century BCE in India.

DEADLY DIAMONDS

Diamonds are highly valued by people all around the world and have been for centuries. They're revered as beautiful, priceless ornaments that decorate royal treasures. They're given as gifts to signify life milestones, such as weddings. However, diamonds are also associated with death.

Natural diamonds are billions of years old, and many famous stones belonged to a variety of owners over the centuries. The Hope Diamond had many royal owners who often met with gruesome ends. At one point, it was among the royal jewels belonging to Louis XVI and Marie Antoinette. The Koh-i-Noor was fought over, stolen, and ultimately signed away in a treaty before becoming part of Queen Victoria's collection in England.

Because diamonds were only found in India prior to the 18th century, their legends and curses often center around ideas of colonialism and theft. Someone stole the Black Orlov diamond from a statue of a Hindu god. Other famous stones were smuggled out of India. In the 1600s, stories say a courier swallowed the Sancy diamond to keep it from robbers. Another story tells of an enslaved person who tried to smuggle the Regent diamond inside an open leg wound.

Diamonds are often associated with tragedies such as the sinking of the *Titanic*. However, most tales aren't true. As you might expect from a ship carrying so many wealthy passengers, valuable pieces of jewelry have been recovered from the debris field. The diamond necklace shown in the 1997 *Titanic* movie wasn't real, but a similar sapphire necklace belonged to a passenger who escaped.

Did you know that a person's ashes can be turned into a diamond? Laboratories can replicate the heat and pressure of centuries in the earth to create a diamond from the carbon remaining in ashes after cremation.

CHAPTER 4

LIFE IS LIKE A SLICE OF PIZZA

Every living thing on earth contains carbon. It's one of the major building blocks of life. People, plants, pets, and even life-forms too tiny to see without a microscope all contain carbon. Carbon makes up just under 20 percent of the human body. If you picture a pizza cut into five big pieces, 20 percent is one-fifth of the pizza, or of those big pieces!

Other main ingredients of life include things like hydrogen, oxygen, nitrogen, sulfur, and phosphorus. Carbon combines easily with all these other elements. Think of all the things that go into a pizza—cheese, tomatoes, maybe pepperoni, and bread. They're all good things on their own, but then a chef puts them together to make something different—a pizza.

Now think about all the ways a cook can make a pizza—different kinds of sauce, cheese, and toppings. All these kinds are still pizzas, but they look and taste different. Like the ingredients that can go on a pizza, carbon atoms connect to each other to make something greater. They form chains that are straight or branching. Other elements can connect to these chains, forming all sorts of new chemical combinations. Deoxyribonucleic acid (DNA) is one of these chains. Every one of your cells contains DNA—it's the blueprint for your unique self.

DNA DETECTIVES

The cells of every life-form on earth—from the anchovies on your pizza to the chef who cooked that pizza—contain DNA. When scientists examine DNA under a microscope, it looks like a long, twisting ladder—two strands connected by rungs. A carbon sugar called deoxyribose helps form the long strands. The "rungs" are formed by pairs of chemicals called bases. DNA contains four different bases: adenine, thymine, guanine, and cytosine. The order in which the pairs are arranged creates the code, or recipe, for a different living being.

Each living thing's DNA is unique. Even the DNA of identical twins isn't quite the same, although it might take an expert to find the differences. This makes DNA a useful tool for helping to solve crimes, sometimes even crimes that happened a long time ago.

In the 1990s, a scientist revealed a grave containing several bodies in Russia. People thought they might be members of the Russian imperial family who were executed in 1918. But how could they be sure after so long? Scientists examined the DNA in the bones. Although each person's DNA is different, there are many similarities between family members. Scientists determined that two of the bodies in the grave were the parents of three of the others.

One of the bodies was thought to be the Empress Alexandra. She was Queen Victoria's granddaughter. Luckily, Queen Victoria had living descendants. One of them, Prince Philip, Duke of Edinburgh, donated a blood sample that allowed scientists to identify one of the bodies in the grave as Alexandra and three of the others, her children. In time, other bodies were also identified as part of the family.

For decades, there were many stories that the Grand Duchess Anastasia (shown here) had somehow survived the execution of her family. However, in 1994, scientists identified her body from the DNA in her bones. The same science was used to disprove the story of Anna Anderson, who claimed to be Anastasia for many years.

THE CHEMISTRY OF LIFE

Knowing that carbon can readily combine with other elements to create DNA, let's look at other combinations, or compounds. Compounds can be made up of just two atoms, such as the simple formula of one carbon atom and one oxygen atom to create carbon monoxide (CO). Or carbon compounds can be made up of several atoms, such as chloroform ($CHCl_3$), which is one atom of carbon, one atom of hydrogen, and three atoms of chlorine (Cl). Some compounds are very complex and can be made up of hundreds of atoms. Because so many elements can interact with carbon, the field of organic chemistry has been developed to study carbon and its compounds. It is called organic chemistry because it is the study of organic compounds, which all contain carbon.

REDUCE, REUSE, RECYCLE

You probably recognize the often-green bins that you can use to recycle bottles and other things, but recycling is not a new concept. Recycling has been part of nature for as long as there's been life on Earth. The natural process of recycling carbon is called the carbon cycle. People and most animals inhale oxygen (O_2) and exhale carbon dioxide (CO_2).

Plants use carbon dioxide exhaled by animals to make the oxygen that animals breathe and the carbohydrates they eat.

Plants are Earth's oxygen factories. NASA is studying how plants might be used to help recycle the air in outer space habitats.

One step in the carbon cycle is the absorption of carbon dioxide by plants as part of photosynthesis. Photosynthesis is when a plant absorbs carbon dioxide from the air, combines it with the water (H_2O) it receives from its roots, and transforms it into oxygen (O_2), a sugar called sucrose ($C_{12}H_{22}O_{11}$), and starch. Starch is mainly sugar molecules strung together. Any of these sugars or starches found in foods, especially in plants like potatoes and wheat, are carbohydrates. Because pastas and breads are made with ingredients from plants like wheat, these foods also contain carbohydrates. Carbohydrates give us energy. When you eat pasta, you are consuming carbohydrates and getting energy from them.

Animals absorb starches and sugars the same way you do—they eat them. When a cow eats grass, it absorbs carbohydrates from the blades of grass. Then the carbon cycle begins anew. Your body transforms carbohydrates into energy to keep your heart pumping and your lungs breathing. When you exhale, carbon dioxide is absorbed by a plant for photosynthesis, and the cycle begins again.

There's more to the carbon cycle too. Animals—including people—go through life breathing oxygen, eating carbohydrates, and exhaling carbon dioxide. However, when a plant or animal dies, its body decays. This process is called decomposition. The body of a dead plant or animal breaks down, releasing carbon. This carbon enriches the soil for more plants to grow and continue the carbon cycle.

The idea of living "carbon neutral" means balancing the amount of carbon you consume and the amount of carbon you return to the planet. Many large companies are working to reduce their amount of carbon pollution.

Over centuries, carbon from decomposed bodies becomes part of the earth. Immense heat and pressure in the earth eventually changes that carbon into fossil fuels, like coal and oil. This is what happened to the dinosaurs, eons ago. When fossil fuels are burned, they release carbon dioxide and other compounds into the air. However, fossil fuels take millions of years to form deep inside the earth. People are burning them faster than the earth can replenish them. Burning too much oil or coal also means too much carbon is released into the atmosphere. This overabundance of carbon compounds in the air is slowly changing the climate.

The carbon cycle works best when all its elements are in balance. A carbon atom can journey from the air to a plant to an animal and back to the air. Then it might start the cycle all over again and end up in the soil. Every living thing plays a role in the carbon cycle.

Burning fossil fuels hurts the earth in two ways: it uses up a finite resource and adds to pollution.

The Paris Agreement is an international treaty aimed at lowering greenhouse gas emissions and encouraging green technologies to help preserve the earth for future generations. As the famous quote says, "We do not inherit the earth from our ancestors, we borrow it from our children."

CHAPTER 5

CALLING ALL COMPOUNDS

Carbon is one of the most common elements on Earth. Its ability to combine with other elements makes it extremely versatile. Carbon compounds form many substances you use every day, including plastic, fuel, and art supplies. Carbon compounds are also found in things you drink, like soda, and medicine you take for an upset stomach. Of course, there are many other substances that contain carbon compounds. Some are useful, and others are harmful.

HYDROCARBONS: HELPFUL OR HARMFUL?

Along with fueling our bodies with carbohydrates, carbon and hydrogen can also team up to provide the energy that runs our cars and provides electricity for our homes. These chemical compounds are called hydrocarbons. There are a number of different molecular combinations of hydrogen and carbon. For instance, methane is a hydrocarbon, and its formula is CH_4. Propane, another hydrocarbon, is C_3H_8. Hydrocarbons are found in petroleum, natural gas, and coal. When you ride in a car, a bus, or an airplane, you are experiencing the energy of carbon. As combustion or heat is applied to the gas, it breaks apart the hydrocarbon molecules. This causes energy to be released to power the car, the bus, or the plane.

However, one of the drawbacks of using fossil fuels, which contain hydrocarbons, is that there is a limited quantity of them. Fossil fuels take millions of years to be converted from ancient animal and plant matter into fuel. Some experts estimate that at our current rate of consumption, Earth's supplies of fossil fuels will run out this century. If we want to continue to travel as freely as we do now with our automobiles, trains, and airplanes, and if we want to have electricity, we have to find a new power source.

One of the more versatile uses of carbon is in the form of dry ice, or frozen carbon dioxide. Dry ice maintains a cold temperature without refrigeration, so it's often used to ship perishable items such as food or medicine. Adding warmer liquid to dry ice changes the solid to a vapor. This is often used as a theatrical effect or in fancy drinks.

"Net zero" emissions means changing the way we live so that we create no more carbon emissions than Earth can easily absorb.

REDUCING OUR CARBON FOOTPRINT

Another downside of burning fossil fuels for energy is that they release pollutants. Carbon dioxide, carbon monoxide (CO), and cyanide are just a few of the pollutants that are expelled from the exhaust pipes of most automobiles and other vehicles that run on fossil fuels. Carbon dioxide is a greenhouse gas. This is because, like the glass panes of a greenhouse, too much carbon dioxide will allow solar radiation into Earth's atmosphere, but it will not let it out. The amount of carbon dioxide and other greenhouse gases being released into the atmosphere has increased greatly ever since the invention of the automobile and other similar machines. The increase of greenhouse gases is a cause of global warming, which has led even the massive polar ice caps to show signs of melting. Greenhouse gases and the greenhouse effect also have been linked to other major environmental concerns. To combat these problems, governments worldwide are encouraging industries and scientists to explore other fuel sources, such as solar energy, wind power, and hydropower. Some automakers are developing cars that run on hydrogen and create virtually no pollution. These alternative cars are starting to gain availability and popularity.

SILENT BUT DEADLY

Some other carbon compounds, especially carbon monoxide, can be a serious threat to your health—and carbon monoxide can be found right inside your home! Carbon monoxide is expelled by home furnaces, kitchen stoves and ovens, and water heaters that burn fossil fuels. When these appliances are working properly, carbon monoxide and other harmful fumes are released through vents outside the home. But if these appliances malfunction, carbon monoxide can seep into the air inside the house and become deadly to its occupants. In fact, more than 200 people in the United States are killed each year by home appliances leaking carbon monoxide. Because carbon monoxide is invisible, odorless, and tasteless and does not make any sound when it is released from a leaky pipe, it has been called a silent killer. Fortunately, inexpensive carbon monoxide detectors can be purchased at almost any hardware store to prevent carbon monoxide poisoning.

Paper and bamboo are good alternatives to some plastic items, such as disposable straws, plates, and utensils. These materials are biodegradable. They break down naturally and don't add to pollution.

PAPER OR PLASTIC?

Fossil fuels are also used to create plastic used for a lot of different things, including cups, pens, and many more items. They are very multifunctional. One of the good things about using plastic is that it will not break like glass. It is also very lightweight.

Plastic is very useful, but it also takes a very long time to decompose in a landfill, the final resting place for much of our garbage. It can take up to 20 years for one plastic fork to begin showing signs of decay. Using fossil fuels to create plastics is not seen as the best way to use our depleting natural resources, and many believe it isn't environmentally responsible. Many fast-food restaurants now use plates and containers made from paper. This paper will decompose faster in a landfill than plastic will.

HOW MANY CAN YOU FIND?

In addition to the numerous carbon-hydrogen combinations, there is an almost never-ending amount of other carbon-based compounds. This can be seen in all the words and phrases containing "carb" or "carbon." For example, the word "carburetor" refers to a component of a car engine that combines fossil fuels and air. Some of the phrases commonly used today that relate to carbon, such as "carbon footprint" and "carbon neutral," only date back to the 1990s.

Plastic takes a long, long time to decompose. When it does start to break down, tiny particles called microplastics are released. Microplastic particles contaminate the water that we drink. Marine animals also ingest microplastic, which can make them sick.

IN YOUR BONES AND IN YOUR HOME

Calcium (Ca) is one of the most common elements on Earth. However, it only exists in nature as part of compounds. One of the most common—and versatile—carbon compounds is formed when calcium and carbon combine. The resulting compound is called calcium carbonate ($CaCO_3$). Calcium carbonate is found many places in the natural world. As a mineral, it's often mined. It's also present in bones, seashells, marble, eggshells, limestone, and coral.

THE WHITE CLIFFS OF DOVER

The white cliffs of Dover, England, are a famous formation of calcium carbonate. The cliffs are made of chalk, a soft form of limestone. Limestone forms from the shells of dead sea creatures. Over time, layers of shells build up on the ocean floor and are covered with mud. Gradually, pressure in the earth's crust hardens these layers into rock. It takes millions of years for limestone to form. Because of this, it often contains fossils of creatures such as sponges and urchins.

The chalk deposits that make up the cliffs in Dover date back to the Cretaceous period (145 million–66 million years ago). In fact, the name "Cretaceous" is taken from the Latin word for chalk. These chalk deposits extend beneath the English Channel and into mainland Europe.

Calcium carbonate is inside your home too. It exists as the plaster in the walls, the chalk in your art set, and the pearls in a jewelry box. Another place is the medicine cabinet. It's the main ingredient in many brands of stomach medicine. Calcium carbonate helps neutralize stomach acid to relieve heartburn and indigestion. It's also found in the vitamin aisle as a dietary supplement. Calcium is necessary for strong, healthy bones and teeth, so some people take vitamin supplements to be sure they're getting enough.

The white cliffs of Dover, England are formed from the skeletons of tiny algae called coccoliths. It's thought that the chalk deposits built up very slowly—as little as half a millimeter a year.

CHAPTER 6

ALL ABOUT ISOTOPES

Carbon occurs in many forms. Aside from all the compounds it forms, it also occurs naturally in three isotopes. An isotope of an element is an atom that has the same number of protons but a different number of neutrons. Scientists study carbon isotopes to learn about the world.

The most common isotope of carbon is carbon-12, which has six neutrons and six protons. Almost all the carbon that occurs in nature—about 99 percent—is carbon-12. Carbon-13 has seven neutrons. About 1 percent of naturally occurring carbon molecules are carbon-13. Carbon-12 and carbon-13 are stable isotopes. This means they don't change forms over time. Carbon-13 is heavier than carbon-12 and occurs in different concentrations in land, air, and sea. Scientists use a device called an isotope ratio mass spectrometer to detect the different concentrations, or isotopic fingerprints.

Carbon-14, also known as radiocarbon, is the rarest natural carbon isotope. Only about one in every trillion carbon molecules is carbon-14! Carbon-14 is the only naturally occurring radioactive isotope. It has a half-life of 5,730 years. This means that it will take 5,730 years for half of this element to decay. Carbon-14 decays into the stable isotope hydrogen-14. Besides being present in nature, carbon-14 can be made artificially in a nuclear reactor. Carbon-14 is present in every organic thing because it combines with oxygen to form CO_2 and then becomes part of the carbon cycle.

Carbon-14 is especially useful in archaeology in a process called carbon dating. Through the food we eat and our daily existence on Earth, living things are constantly taking in carbon-14. Nearly everything alive is absorbing carbon-14. But when you die, you stop absorbing carbon-14, so the present amount in your body slowly begins to decay. Archaeologists can guess the age of a fossil by how much carbon-14 is found in the specimen.

So, when a scientist measures the amount of carbon-14 present in a fossil, they compare it with a plant or animal in its present-day form. For example, scientists can compare the amount of carbon-14 in a fossilized fern leaf with the amount of carbon-14 in a modern fern leaf. From this, scientists can estimate when the fossilized fern died by figuring out when it stopped producing carbon-14.

The oldest ferns in the fossil record date back 360 million years.

Scientists who study the atmosphere use radiocarbon dating to determine what percentage of airborne pollution comes from fossil fuels.

So, let's say you are studying to be an archaeologist and are digging at a site where ancient people may have dwelled. While you are digging, you come across two bones that look to be from an animal, possibly a deer. To determine if these bones came from a deer that had been dead for a few years or for thousands of years, you will take it back to the lab and measure the carbon-14 levels in each bone. Maybe the carbon-14 levels are really high in one of the bones. In fact, it is the same amount of carbon-14 as the levels in a deer that is only a few days old. Meanwhile, the other bone you found has very low levels of carbon-14. This bone has levels consistent with other bones that have been determined to be from the year 1200. Knowing this, you can estimate that the first bone you found is from a recently deceased deer, and the second bone is from a deer more than 800 years old.

The Shroud of Turin was revered for centuries as the burial cloth of Jesus Christ. Carbon dating that took place at three separate labs in the 1980s dated the cloth to the Middle Ages. However, new investigations suggest that the fibers tested were from medieval cloth used to repair the shroud. Answering old questions with new technology is one of the ways we learn about the world.

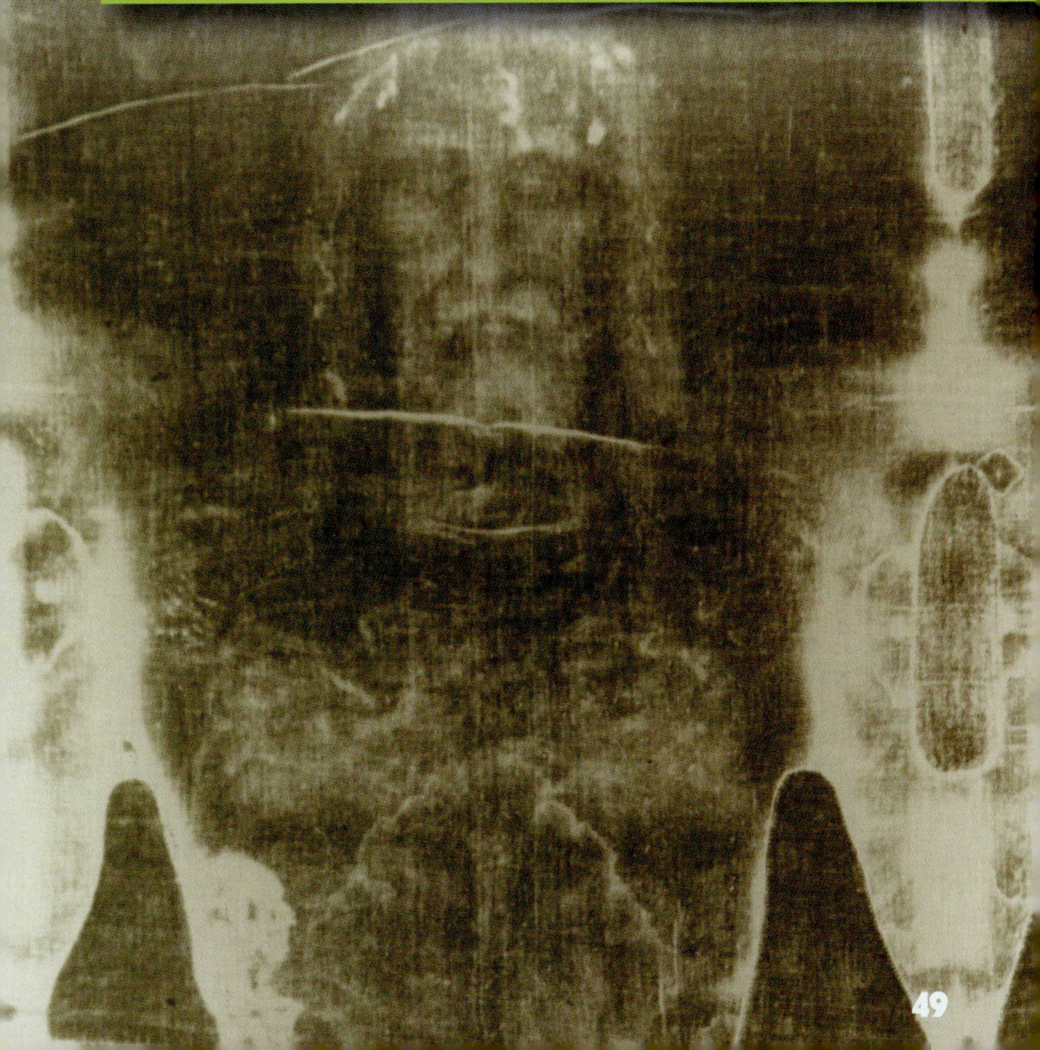

In order for new elements to be recognized and added to the periodic table, teams of scientists in different labs around the world need to be able recreate the same results.

A COMPOUND BY ANY OTHER NAME...

The naming of organic compounds is regulated by the International Union of Pure and Applied Chemistry (IUPAC). The goal is for chemists all over the world to use the same system of nomenclature, or style of naming things. This makes scientific cooperation between chemists from different countries easier. If everyone uses the same format, then scientists from different places know that they're talking about the same substance, even if it has different local names.

Many words used in science have Latin origins. In addition, compounds might be named for the types of bonds they have: an alkane is a single-bonded hydrocarbon, an alkene is double bonded, and a triple bond is an alkyne. Other factors might include whether it has a cyclic structure (the atoms are arranged in a ring), or if the compound includes hydrogen.

Using the IUPAC system of naming carbon compounds involves using a "root word" based on the number of carbon atoms the compound contains.

NUMBER OF CARBON ATOMS	ROOT WORD
ONE	METHYL
TWO	ETHYL
THREE	PROPYL
FOUR	BUTYL
FIVE	PENTA
SIX	HEXA

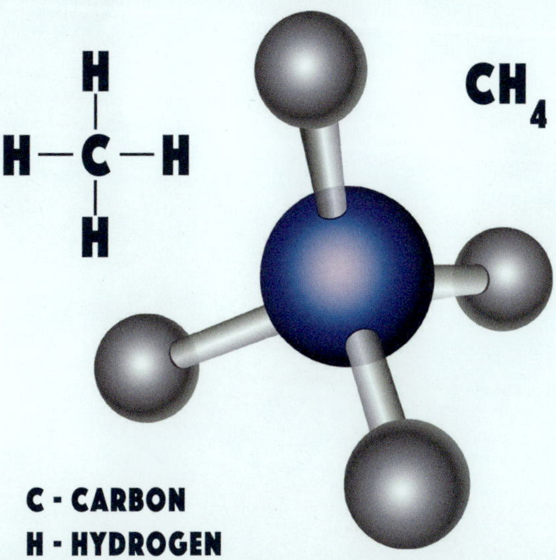

Methane (shown here) is a greenhouse gas. It's a simple hydrocarbon formed from one carbon atom (hence the root word methyl) and four hydrogen atoms.

Carbon is all around us. It's in the food we eat, the fuel that powers our vehicles, and walls of our homes. You use carbon in your classroom when you write with chalk or a pencil. Carbon even makes up the DNA that defines who we are.

Carbon can exist as either an organic or an inorganic compound. It can also exist as a metallic compound. Carbon is a huge part of the never-ending cycle of life on earth.

Your carbon footprint is the amount of greenhouse gases released into the atmosphere as a result of your actions. What can you do to reduce yours?

There are limited amounts of fossil fuels left in the earth, so cities of the future will rely more on clean energy sources like wind and solar power.

THE PERIODIC TABLE OF ELEMENTS

Carbon is tetravalent. This means it has four valence electrons and can form up to four covalent bonds with other elements. Carbon has more compounds than any other element.

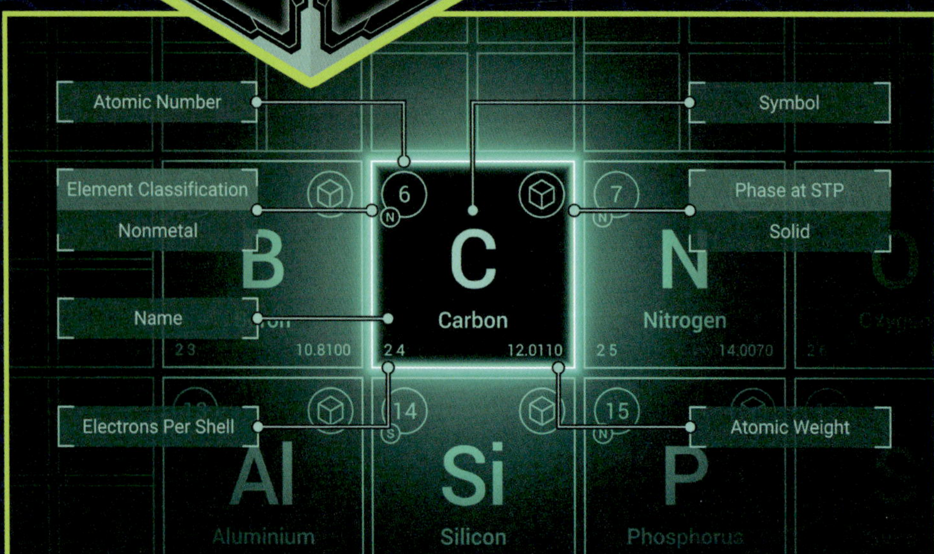

PERIODIC TABLE OF ELEMENTS

- Alkali metals
- Alkaline earth metals
- Transition metals
- Metal
- Metalloids
- Nonmetals
- Halogen
- Noble gases
- Lanthanoids
- Actinoids

1 H Hydrogen 1.008																	2 He Helium 4.0026
3 Li Lithium 6.94	4 Be Beryllium 9.0122											5 B Boron 10.81	6 C Carbon 12.011	7 N Nitrogen 14.007	8 O Oxygen 15.999	9 F Fluorine 18.998	10 Ne Neon 20.180
11 Na Sodium 22.990	12 Mg Magnesium 24.305											13 Al Aluminium 26.982	14 Si Silicon 28.085	15 P Phosphorus 30.974	16 S Sulfur 32.06	17 Cl Chlorine 35.45	18 Ar Argon 39.948
19 K Potassium 39.098	20 Ca Calcium 40.078	21 Sc Scandium 44.956	22 Ti Titanium 47.867	23 V Vanadium 50.942	24 Cr Chromium 51.996	25 Mn Manganese 54.938	26 Fe Iron 55.845	27 Co Cobalt 58.933	28 Ni Nickel 58.693	29 Cu Copper 63.546	30 Zn Zinc 65.38	31 Ga Gallium 69.723	32 Ge Germanium 72.630	33 As Arsenic 74.922	34 Se Selenium 78.971	35 Br Bromine 79.904	36 Kr Krypton 83.798
37 Rb Rubidium 85.468	38 Sr Strontium 87.62	39 Y Yttrium 88.906	40 Zr Zirconium 91.224	41 Nb Niobium 92.906	42 Mo Molybdenum 95.95	43 Tc Technetium (98)	44 Ru Ruthenium 101.07	45 Rh Rhodium 102.91	46 Pd Palladium 106.42	47 Ag Silver 107.87	48 Cd Cadmium 112.41	49 In Indium 114.82	50 Sn Tin 118.71	51 Sb Antimony 121.76	52 Te Tellurium 127.60	53 I Iodine 126.90	54 Xe Xenon 131.29
55 Cs Caesium 132.91	56 Ba Barium 137.33	57-71	72 Hf Hafnium 178.49	73 Ta Tantalum 180.95	74 W Tungsten 183.84	75 Re Rhenium 186.21	76 Os Osmium 190.23	77 Ir Iridium 192.22	78 Pt Platinum 195.08	79 Au Gold 196.97	80 Hg Mercury 200.59	81 Tl Thallium 204.38	82 Pb Lead 207.2	83 Bi Bismuth 208.98	84 Po Polonium (209)	85 At Astatine (210)	86 Rn Radon (222)
87 Fr Francium (223)	88 Ra Radium (226)	89-103	104 Rf Rutherfordium (267)	105 Db Dubnium (268)	106 Sg Seaborgium (269)	107 Bh Bohrium (270)	108 Hs Hassium (277)	109 Mt Meitnerium (278)	110 Ds Darmstadtium (281)	111 Rg Roentgenium (282)	112 Cn Copernicium (285)	113 Nh Nihonium (286)	114 Fl Flerovium (289)	115 Mc Moscovium (290)	116 Lv Livermorium (293)	117 Ts Tennessine (294)	118 Og Oganesson (294)

57 La Lanthanum 138.91	58 Ce Cerium 140.12	59 Pr Praseodymium 140.91	60 Nd Neodymium 144.24	61 Pm Promethium (145)	62 Sm Samarium 150.36	63 Eu Europium 151.96	64 Gd Gadolinium 157.25	65 Tb Terbium 158.93	66 Dy Dysprosium 162.50	67 Ho Holmium 164.93	68 Er Erbium 167.26	69 Tm Thulium 168.93	70 Yb Ytterbium 173.05	71 Lu Lutetium 174.97
89 Ac Actinium (227)	90 Th Thorium 232.04	91 Pa Protactinium 231.04	92 U Uranium 238.03	93 Np Neptunium (237)	94 Pu Plutonium (244)	95 Am Americium (243)	96 Cm Curium (247)	97 Bk Berkelium (247)	98 Cf Californium (251)	99 Es Einsteinium (252)	100 Fm Fermium (257)	101 Md Mendelevium (258)	102 No Nobelium (259)	103 Lr Lawrencium (266)

- C Solid
- Hg Liquid
- H Gas
- Rf Unknown

GLOSSARY

allotrope: A substance and especially an element in two or more different forms usually in the same phase.

atom: The smallest, most basic unit of an element. Atoms can exist alone or in combinations.

atomic mass: The mass of one atom of an element.

atomic number: The number of protons in the nucleus of an atom, which is equal to the number of electrons. The atomic number determines an element's place on the periodic table.

combustion: The act of burning; a chemical process that produces heat and light.

compound: Something formed by a union of elements or parts.

decay: The spontaneous decrease in the number of radioactive atoms in radioactive material.

decomposition: The act of breaking down into simpler parts or elements.

electron: A negatively charged particle found outside the nucleus of an atom.

element: The basic matter that all things are made of; any substance made of one kind of atom.

group: The elements that are in a column of the periodic table.

hydrocarbon: An organic compound (such as acetylene or butane) containing only carbon and hydrogen and often occurring in petroleum, natural gas, coal, and bitumens.

isotope: Any of two or more species of atoms of a chemical element with the same atomic number and nearly identical chemical behavior but with differing atomic masses or mass numbers and different physical properties.

molecule: The smallest particle of a substance.

neutron: A particle within the nucleus of an atom that contains no charge.

nomenclature: A system or set of terms or symbols especially in a particular science, discipline, or art.

nucleus: The positively charged central portion of an atom.

period: A horizontal row in the periodic table.

proton: A positively charged particle within the nucleus of an atom. The number of protons and electrons are almost always equal in an atom.

radioactive: Of, caused by, or showing radioactivity, the property possessed by some elements (such as uranium) or isotopes (such as carbon-14) of spontaneously emitting energetic particles (such as electrons or alpha particles) by the disintegration of their atomic nuclei.

subatomic: Of, relating to, or being particles smaller than atoms.

FOR MORE INFORMATION

AMERICAN MUSEUM OF NATURAL HISTORY

200 Central Park West
New York, NY 10024-5102
(212) 769-5000
Website: www.amnh.org
Facebook: @naturalhistory
Instagram and X: @amnh
This is a world-class museum dedicated to sharing scientific knowledge about the natural world.

CHEMICAL INSTITUTE OF CANADA (CIC)

90-2420 Bank Street
Ottawa, Ontario, Canada
K1V 8S1
(613) 232-6252
Website: www.cheminst.ca/
X: @CIC_ChemInst
The CIC facilitates the research and cooperation of chemists and chemical engineers in Canada.

INTERNATIONAL UNION OF PURE AND APPLIED CHEMISTRY (IUPAC)

IUPAC Secretariat
79 T.W. Alexander Drive
Research Commons Building 4501, Suite 190
Research Triangle Park, NC 27709, USA
(919) 485-8700
Website: iupac.org/
Facebook: @iupac.org
X: @iupac
IUPAC is the global chemistry organization that maintains the periodic table of elements.

INTERNATIONAL UNION OF PURE AND APPLIED PHYSICS (IUPAP)

Fondazione Internazionale Trieste per il Progresso e la Libertà delle Scienze
c/o ICTP
Strada Costiera 11
34151 Trieste
Website: iupap.org/
Facebook: @iupap
X: @iupap_physics
IUPAP's mission is to assist in the global advancement of physics and facilitate international scientific cooperation.

LOS ALAMOS NATIONAL LABORATORY

PO Box 1663
Los Alamos, NM 87545
(505) 667-5061
Website: www.lanl.gov/
Facebook: @LosAlamosNationalLab
Instagram and X: @losalamosnatlab
Los Alamos National Laboratory is a federally funded research and development center.

FOR FURTHER READING

Carbon Almanac Network. *Generation Carbon: It's Time to Start: A Carbon Almanac for Kids*. Independently published, 2022.

Congdon, Lisa. *The Illustrated Encyclopedia of the Elements: The Powers, Uses, and Histories of Every Atom in the Universe*. San Fransico, CA: Chronicle Books LLC, 2021.

DK. Super Simple Chemistry: *The Ultimate Bitesize Study Guide*. New York, NY: DK Publishing, 2020.

Emminizer, Theresa. *The Carbon Cycle*. Buffalo, NY: Enslow Publishing, 2024.

McHenry, Ellen Johnston. *The Chemical Elements Coloring & Activity Book: A Fun and Interactive Guide to the World of Chemistry*. Independently published, 2021.

Silver, Donald, and Patricia Wynne. *My First Book About Chemistry*. Mineola, NY: Dover Publications, 2020.

Thomas, Isabel, and Sara Gillingham. *Exploring the Elements: A Complete Guide to the Periodic Table*. New York, NY: Phaidon Press Inc., 2021.

Zovinka, Edward P., and Rose A. Clark. *A Kids' Guide to the Periodic Table: Everything You Need to Know About the Elements*. Emeryville, CA: Rockridge Press, 2020.

INDEX

A

allotropes, 22–23, 26
amorphous carbon, 27
atomic number, 9, 11, 13, 15, 20
atomic weight/mass, 8–9, 11, 13–14

B

Boyle, Robert, 6–7
buckminsterfullerene, 11, 26

C

calcium, 44–45
calcium carbonate, 44–45
carbohydrates, 33–35, 38
carbon cycle, 5, 33–36, 47
carbon dating, 47–49
carbon dioxide, 5, 12, 16, 33–36, 39–40
carbon group, 17
carbon monoxide, 32, 40–41
charcoal, 7, 11, 27
chlorine, 32
chloroform, 32
coal, 7, 27, 36, 39
combustion, 5, 39
covalent bonds, 12, 16, 18
crystal, 22–24, 27
cyanide, 40

D

Davy, Humphry, 7
deoxyribose, 31
diamond, 4, 7–8, 11, 18, 22–26, 28–29
DNA, 31–32, 52

E

electricity, 16, 24, 38–39
electrons, 9–12, 16–18, 20, 22

F

flerovium, 17, 20
fossil fuels, 27, 36, 39–40, 42, 48, 53
fullerenes, 23, 26

G

gases, 4–5
germanium, 17
graphite, 5, 8, 11, 18, 22–26

H

heat, 16, 23–24, 27, 29, 39
hydrocarbons, 38–39, 50–51
hydrogen, 14, 30, 32, 38, 41, 43, 47, 50–52

I
International Union of Pure and Applied Chemistry (IUPAC), 20, 50
ionic bonds, 12
iron, 8, 27
isotopes, 20, 46–49

L
Lavoisier, Antoine, 7
lead, 18

M
magnets, 17
Mendeleev, Dmitri, 14–15
metals, 17
methane, 38, 51

N
net zero, 40
neutrons, 8–11, 17, 20, 46
nitrogen, 30
nucleus, 9–10, 22

O
organic chemistry, 32–33
oxygen, 12, 18, 30, 32–35, 47

P
periodic table, 8, 10, 13–17, 19, 50
phosphorus, 30
photosynthesis, 5, 34–35
plastic, 38, 42–43
potassium, 7
propane, 38
protons, 8–11, 17, 20, 46

R
radiocarbon, 46, 48
representative elements, 16
respiration, 5
Rutherford, Ernest, 9

S
silicon, 17
sodium, 7
solids, 4
soot, 7, 27
sugars, 31, 34–35
starch, 34–35
sulfur, 30

T
tin, 17
transition elements, 16

U
unit cell, 22
uranium, 14, 20

V
valence shells, 22

W
white carbon (chaoite), 11, 26

ABOUT THE AUTHOR

KATHLEEN A. KLATTE is the author of many nonfiction books for children and teens. Topics she has written about range from animals and nature to constitutional law to unusual career choices. She lives in New York with one cat and far too many books and Legos.

PHOTO CREDITS

Cover, pp. 1, 20 remotevfx.com/Shutterstock.com; Cover, pp.1-64 pluie_r/Shutterstock.com; Cover, p. 3-64 Oksancia/Shutterstock.com; p. 4 pedrosek/Shutterstock.com; p. 5 Designua/Shutterstock.com; p. 7 https://en.m.wikipedia.org/wiki/File:The_Shannon_Portrait_of_the_Hon_Robert_Boyle.jpg; p. 8 vchal/Shutterstock.com; p. 9 saran insawat/Shutterstock.com; p. 10 zizou7/Shutterstock.com; p. 11 tunasalmon/Shutterstock.com; p. 12 Sansanorth/Shutterstock.com; p. 13 JLStock/Shutterstock.com; p. 15 Srg Gushchin/Shutterstock.com; p. 16 Anterovium/Shutterstock.com; p. 17 Veli_G/Shutterstock.com; p. 18 Icruci/Shutterstock.com; p. 19 Gorodenkoff/Shutterstock.com; p. 21 Mikhail Markovskiy/Shutterstock.com; p. 23 Abdullaharsln/Shutterstock.com; p. 24 DiamondGalaxy/Shutterstock.com; p.25 Akifyeva S/Shutterstock.com; p. 26 BhuvichRyu/Shutterstock.com; p. 27 Vector FX/Shutterstock.com; p. 28 onair/Shutterstock.com; p. 29 Retouch man/Shutterstock.com; p. 32 Everett Collection/Shutterstock.com; p. 33 Proxima Studio/Shutterstock.com; p. 34 chayanuphol/Shutterstock.com; p. 35 wee dezign/Shutterstock.com; p. 36 Ha-nu-man/Shutterstock.com; p. 37 BOY ANTHONY/Shutterstock.com; p. 39 w583254846/Shutterstock.com; p. 40 Igor Starodubtsev/Shutterstock.com; p. 42 Tetiana Shumbasova/Shutterstock.com; p. 43 Siberian Art/Shutterstock.com; p. 45 Konmac/Shutterstock.com; p. 47 Jirik V/Shutterstock.com; p. 48 Kristen Badali/Shutterstock.com; p. 49 godongphoto/Shutterstock.com; p. 50 greenbutterfly/Shutterstock.com; p. 51 Radzas2008/Shutterstock.com; p. 52 Sansoen Saengsakaorat/Shutterstock.com; p. 53 Owlie Productions/Shutterstock.com; p. 54 ShannonChocolate/Shutterstock.com; p. 55 View_From_My_Lens/Shutterstock.com.